Daiana Oliveira Reis
João Alberto Santos Porto
Francesca Regina R. Neto

An Analysis of Heat Islands in the City of Caxias Maranhão

Daiana Oliveira Reis
João Alberto Santos Porto
Francesca Regina R. Neto

An Analysis of Heat Islands in the City of Caxias Maranhão

The influences of urbanism on climate change

ScienciaScripts

Cover image: www.ingimage.com

This book is a translation from the original published under ISBN 978-3-330-75614-4.

Publisher:
Sciencia Scripts
is a trademark of
Dodo Books Indian Ocean Ltd. and OmniScriptum S.R.L publishing group

120 High Road, East Finchley, London, N2 9ED, United Kingdom
Str. Armeneasca 28/1, office 1, Chisinau MD-2012, Republic of Moldova, Europe
Printed at: see last page
ISBN: 978-620-8-27591-4

SUMMARY

SUMMARY

One of the major problems in cities is the difficulty in obtaining pleasant thermal sensations, which has led the population to resort to artificial forms of heating/cooling. The main objectives of this work are to describe the heat island phenomenon in the urban area of Caxias, Maranhão, to measure the temperature in the morning and afternoon, to compare the temperatures in the center of Caxias with outlying neighborhoods and to present solutions for mitigating it. We believe that this study is necessary, since the consequences of the occupation and development process in Caxias Maranhão, its poor urban planning, large-scale production of waste (solid, liquid and gaseous) have been causing thermal imbalance, hindering evaporation, reducing the dispersion power of atmospheric pollutants, generated mainly in the city center, which has raised its temperature. Mitigating heat islands would bring several benefits, among them: improved health, lower temperatures, balanced relative humidity, energy savings and, consequently, a better quality of life for people.

Key words: Heat islands. Caxias. Temperature. Relative humidity. Urban Planning. Mitigation.

INTRODUCTION

With the development of cities, industries and the widespread use of cars, there has been a continuous increase in the concentration of air pollutants such as carbon dioxide and other greenhouse gases. Among the most significant climate changes resulting from urbanization is the heat island phenomenon. Jensen (2009) attributes the emergence of heat islands to deforestation and the replacement of the ground surface with non-evaporative and non-porous materials, such as asphalt and concrete, which causes rainwater to evaporate more quickly, reducing cooling.

The name urban heat island comes from the fact that a city has a very high heat flow in its center, while in its surroundings it is much higher than in the city center (LOMBARDO, 1985).

Heat islands don't just cause minor additional discomfort; their higher temperatures, lack of shade and role in increasing air pollution have serious effects on mortality and the health of the population. They waste money by increasing energy demand, as well as being unhealthy for urban fauna and flora.

A classic example of a problem typical of cities is the difficulty of obtaining pleasant thermal sensations, which has led the population to resort to artificial forms of heating/cooling, generating an increase in the demand for electricity. In Brazil, the study of thermal comfort by Urban Climatology has confirmed this and other issues in cities with different climatic configurations (SOUZA, 2013; SOUZA *et al.*, 2012).

Planet Earth is getting worse and as a result the average global temperature is rising and increasingly bringing catastrophic results such as: floods, melting polar ice caps, temperature changes, unseasonable droughts and the destruction of the ozone layer.

Caxias is our research area. It occupies a position close to the equatorial belt and, due to its low altitude, is mainly influenced by the hot and humid continental equatorial masses (cEm). It has megathermal characteristics, meaning that the average temperature of the coldest months is over 18 °C. The increase in construction, motor vehicles and the suppression of vegetation (characteristic of urban areas) have raised the local temperature.

Several studies in this area have been carried out, predicting the behavior of nature in all this, and choosing more suitable alternatives so that urban development can take place without major damage to the quality of life. The heat island can generally be controlled by various elements, including the location of the city (water bodies, nature of the soil, vegetation, land use, architecture, building materials and anthropogenic sources), geographical location, topography, urban climatology, seasonality and synoptic weather conditions (OKE, 1982; MOREIRA *et al.*, 2011).

The main objectives of this work are to describe the heat island phenomenon in the urban area of Caxias Maranhão, comparing the temperatures in the center of Caxias with some outlying neighborhoods, measuring the temperature in the morning and afternoon.

We believe that this study is necessary, since the consequences of the occupation and development process in Caxias Maranhão, its poor urban planning, the production of more waste (solid, liquid and gaseous) than the environment can assimilate, have been causing thermal imbalance and these conditions hinder evaporation, reducing the dispersion power of the atmospheric pollutants generated mainly in the city center, thus increasing its temperature.

Based on studies by: GARTLAND, 2010; MONTEIRO; BIAS *et al.*, 2003; COSTA *et al.*, 2009; IWAI, 2003; COSTA *et al.*, 2009; MOREIRA *et al.*, 2011; SOUZA, 2013; SOUZA *et al.*, 2012; TAHA, 1997; VENTURINI, 2009; OLIVEIRA 2012; COSTA *et al.*, 2010; LOMBARDO, 1985.

In June, two teams carried out a heat island analysis in the city of Caxias,

Maranhão, using two thermohygrometers and two laser thermometers, in the morning and afternoon, always within a period of one hour, in nine neighborhoods of the city: Salobro, Campo de Belém, Parque da Cidade, Centro, Volta Redonda, Cohab, Praça da Chapada and the Shopping Mall area, where the ambient temperature, relative humidity and soil temperature were measured within an hour, in the morning and afternoon shifts, on Saturdays, Sundays and Mondays.

In the morning, the measurements started at 9.5 a.m. and ended at 9.45 a.m., and in the afternoon they started at 2.5 p.m. and ended at 2.45 p.m., always with a 10-minute interval from one neighborhood to the next.

The nine points were strategically chosen and observed using Google Earth, where the spatialization of the city was taken into account. We tried to choose points where we could expand, and at the same time cover the city of Caxias as much as possible. The technique used was Mobile Transects, which involves traveling a predetermined route through a region, stopping at representative locations to take measurements using only one type of meteorological instrument.

The work was divided into four chapters, where we talk a little about the city of Caxias, factors that influence the existence of heat islands, we will see the various techniques that can be used to investigate this phenomenon, statistical analysis of the data collected and alternatives as solutions for the mitigation of this meteorological phenomenon that is the Heat Island.

1 LITERATURE REVIEW

1.1 History, Origin and Evolution

Studies on the quality of urban life date back to the Industrial Revolution, although it was in the 20th century, especially since the Second World War, that they developed most strongly. In the last thirty years or so, this issue has become a major concern for scientists, politicians, administrators and society in general due to the considerable changes introduced by the urbanization process in the previous environment and the creation, in most cases, of environmental conditions that are harmful to life (SANTOS, 1991:43; MONTEIRO, 2003).

The first documentation of an urban heat island was in 1818, when Luke Howard's revolutionary study of London's climate detected an "excess of artificial heat" in the city compared to the countryside (Howard 1833). Emilen Renou made similar discoveries about Paris during the second half of the 19th century (Renou, 1855, 1862, 1868) and Wilhelm Schimidt found such conditions in Vienna at the beginning of the 20th century (Scmidt, 1917, 1929). In the United States, studies on heat islands began in the first half of the 20th century (Mitchell, 1953, 1961).

In Brazil, references include Tavares (1975), who studied the local climate of Campinas-SP; Monteiro (1976), who developed a whole theoretical framework for urban climate studies; Sartore (1979), responsible for the first urban climate study of medium-sized cities (Santa Maria- RS); and Lombardo (1985), who used the theoretical framework developed by Monteiro to investigate the phenomenon of heat islands in the metropolis of São Paulo, developing a whole methodological framework using data collected by orbital sensors. (NASCIMENTO, 2011).

According to TARIFA, 1977, in an urban study of Sâo José dos Campos based on a new comparative analysis of the temperature and humidity of the urban and rural areas of the city using only two observation points.

Concerns about the quality of life in the urban environment have been growing more and more, and what Monteiro had launched in 1975 as a theoretical proposal for conducting research into the climates of Brazilian cities was presented as a thesis for an academic degree at the University of São Paulo (USP), in the Department of Geography of the Faculty of Philosophy, Letters and Human Sciences (FFLCH), became a reality: a subject called "Urban Climatology" was introduced into the curriculum of the post-graduate course in Physical Geography.

All these studies have shown how a city's climate is the result of energy transformations and the interaction between the air and the built environment. This finding directly linked human interventions in the environment and their effects, and led to climatic differences between urban and rural environments.

1.2 Urban Development

Humanity's choice to live collectively has made the city famous for providing its inhabitants with all the advantages of access to housing, employment, supplies, energy, sanitation, transport, health, education and leisure. However, alongside these facilities, one of the negative consequences of this urban habitat has undoubtedly been the environmental impacts it generates (MONTEIRO *et al.*, 2003).

As cities expand, heat islands also tend to become more intense. Analyses of historical weather data show that the intensification of heat islands coincides with the development of urban and suburban areas.

A metropolis without adequate land use planning, with a lack of appropriate verticalization and occupation parameters, especially where it grows at a rapid rate and with few technical resources, can put the quality of life of its inhabitants at risk (LOMBARDO, 1985; BIAS *et al.*, 2003).

Urban areas have structural characteristics that make their study complex, as they lack continuity or homogeneity. They are areas made up of various materials that reflect or radiate electromagnetic energy in different ways, presenting a large spectral amplitude and high spatial frequencies that are difficult to separate (COSTA et al., 2009).

Studies carried out throughout the 20th century have determined that urban surfaces are warmer than rural ones: Firstly, man-made surfaces are made up of dark materials that readily absorb and store the sun's heat. Secondly, most building materials are resistant to water which, when it comes into contact with the surface, begins to "run" and is unable to dissipate the heat through evaporation.

The environmental problems caused by the shift from a predominantly rural to an urban population began in the mid-1960s and accelerated considerably in the following decades. These Brazilian metropolitan areas today have commonalities, that is, they are the result of multiple and extremely interrelated issues, among which accelerated demographic growth, the housing problem, the lack of basic infrastructure and socio-economic inequalities have a very strong impact on the natural environment.

CARVALHO (2013) points out that the accelerated process of urbanization that is sweeping across Brazil reflects the transformations that are taking place in the urban space, especially with regard to the growth of built space in large cities. These changes, which bring comfort and well-being to society, are also causing major disruption to the natural environment.

Climate is one of the dimensions of the urban environment and its study has made important contributions to solving environmental issues in cities. The climatic conditions of these areas, understood as urban climate, are derived from the alteration of the natural landscape and its replacement by a built environment, the scene of intense human activities.

However, the act of planning, or guiding the urban development of a locality, is a very recent activity from the perspective of the action of institutions linked to city management in the tropical world. It began in the middle of the last century in some European countries, notably England, France and Germany, and gained full employment after the Second World War. Curitiba, in Brazil, is one of the few examples of a city with good quality green areas in relation to the number of inhabitants.

It is in large cities and metropolitan areas that the greatest number of studies on urban environmental issues can be found, as these are the places where the quality of life is most severely degraded. Small and medium-sized cities also have deteriorated environments, but the study of these is only slowly gaining importance nowadays.

Many studies have already verified the occurrence of heat islands in cities of different sizes, increasing concern about minimizing their effects on the areas they occupy and the population they cover. They constitute "an 'inverse oasis', where air and surface temperatures are warmer than in surrounding rural areas" (GARTLAND, 2010).

A considerable number of small, medium and large cities in Brazil have been the target of research into their urban atmospheric environments since the 1960s (Monteiro, 2003), which intensified and spread to almost the entire country in the 1990s.

From a conceptual point of view, however, there is a significant lack of basic bibliography for the specific treatment of small and medium-sized cities, a fact that takes the framework of concepts produced to date to a dimension of high relativity, since they were produced taking into account, above all, large metropolitan urban agglomerations.

1.2.1 Sustainability and planning

The word city, according to the dictionary, means a human agglomeration located in a circumscribed geographical area and with numerous houses, close together, intended for housing and/or cultural, commercial, industrial, financial and other activities not related to the direct exploitation of the land; urbe.

From this meaning it is clear how the city adapts to the needs created by us, how we interfere with its original image in order to achieve a place that, under our concepts, is more suitable for our activities and our coexistence. Today, the urban city is a mirror of our desires.

The basis of what is defined today as sick cities, or biocides, is undoubtedly disordered urban growth. This term, widely used by biologists, architects, sociologists, environmentalists and even politicians, is justified by the way its effects act on the environment (VENTURINI, 2009).

This condition is a reflection of cities that absorb a large number of activities, which brings with it an increase in urbanization, traffic, construction and consequently the generation of waste.

It should be emphasized that urban growth alone is not and should not be considered the cause of the existence of biocidal cities.

The complexity of urban centers fundamentally characterizes such an existence, which is often treated negligently, where they are almost always presented in inefficient ways to solve the common problems that are frequent.

In the study of urban spaces, (ROMERO, 2004; VENTURINI, 2009) establishes the existence of four guidelines that must be considered when proposing a sustainable project:

> **Linkage:** economic development, affordable housing, public safety, environmental protection and mobility are interrelated and must be addressed in an integrated manner.

> **Inclusion**: a variety of stakeholders must be brought together to identify and achieve common values and objectives.

> **Forecasting**: long-term objectives should be drawn up.

> **Quality**: elements that contribute to maintaining diversity should be sought out and prioritized. This ensures the quality and not just the quantity of spaces, increasing the quality of urban life.

These indices only confirm how environmental, social, political and economic aspects are related and establish an interdependence in the creation and maintenance of sustainable urban spaces.

1.3 The City of Caxias

Caxias, like all cities that originated from colonial settlements, was formed through violent processes and intricate power plays. The embryo of the formation of Caxias can be found in the colonial period. The origin of the settlement is usually attributed to the movements of the entradas and the sertanistas, from which the village of Caxias das Aldeias Altas and, later, the city of Caxias were formed.

Caxias then began to form at the end of the 1720s, after the appeasement of the middle Itapecuru by the successive incursions of bandeirantes, especially from Vila da Mocha, in Piaui, under the leadership of Bernardo Carvalho de Aguiar.

In the 1730s, the town of Caxias was elevated to the category of parish, and the nuclei of the two ongoing colonization projects controlled by the Jesuits were already established, where the converted Indians stayed, working in their fields and receiving catechism and Jesuit education. On the right bank were the crown's exploratory colonizing project, rice plantations, sugar cane, cotton and cattle breeding. A chapel to St. Joseph was built on this bank of the river. These projects were complementary and antagonistic depending on the moment. Later, the Jesuits were expelled from Maranhão, and the settlers' project took over the entire territory.

1.3.1 Atmospheric phenomena

> *Climate*

Maranhão is in a climate transition region, with some areas located close to the Northeastern Semi-Arid (hot and dry), the Amazon (hot and humid), the Atlantic Ocean and the Brazilian Central Plateau. Thus, in Maranhão, there is significant spatial variability in climatological elements such as rainfall, relative humidity and temperature.

There are no well-defined seasons of the year, and only two periods can be considered: a dry period and a rainy period, which in turn are not formed at the state level in homogeneous periods; because from the extreme north of the state up to 5° South Latitude, the formation of the rainy season depends mainly on ideal temperature conditions in the tropical Atlantic Ocean, which interferes with the permanence of the Convergence Zone.

> *Average relative humidity*

Figure 1 - Map of relative humidity in the municipality of Caxias

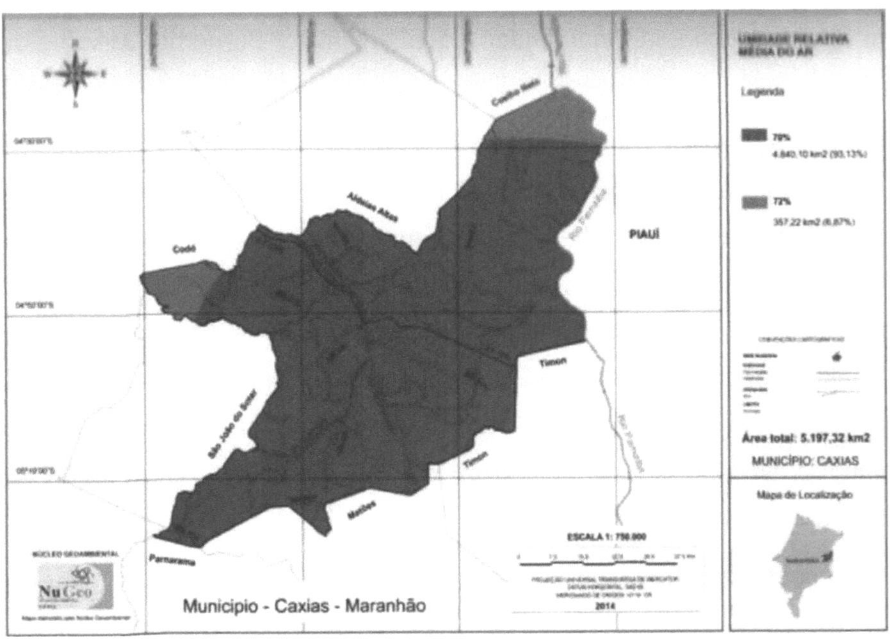

Source: LETRAS, 2015

Relative humidity is the water, in the form of vapor, that exists in the atmosphere. Water vapor (from the evaporation of water from land surfaces, lakes, oceans and plant transpiration) is the decisive element in the hydrological cycle, either transferring water from the surface to the atmosphere or returning in liquid form as rain.

As we can see from the map above, Caxias has an average relative humidity of 70%, and the data collected in the research area will compare this average with those presented in the neighborhoods surveyed.

> *Average annual temperature*

Figure 2 - Map of the average annual temperature in the municipality of Caxias

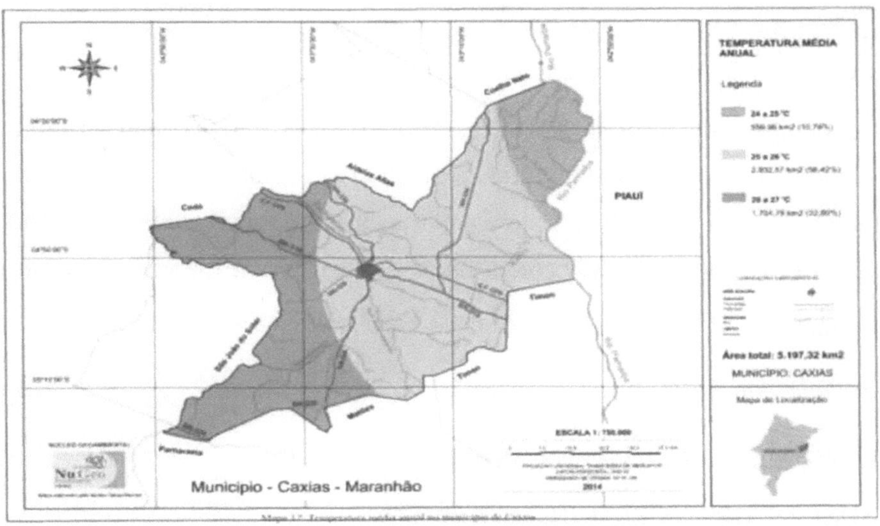

Source: LETRAS, 2015

Air temperature plays a role in the evapotranspiration process, because the solar radiation absorbed by the atmosphere and the heat emitted by the cultivated surface raises the temperature. As we can see from the map above, the area where Caxias is located has an average annual temperature of 25 °C to 26 °C. This data will be compared with the field research carried out.

> *From 1950 to the 21st century - A statistical portrait of Caxias*

In 1950, 107,347 people were enumerated, 53,559 men (51.82%) and 53,778 women (50.11%). 47,197 men (88.12%) and 45,705 women (84.97%) lived in rural areas. It can therefore be seen that 92,902 inhabitants, corresponding to 86.54% of the population, lived in rural areas. The urban area therefore concentrated 13.46% of the population.

Table 1 - Caxiense population in the 1950s

POPULATION

Color	**Men**	**%**	**Women**	**%**	**TOTAL**	**%**
White	8.378	15,67	8.803	16,39	17.181	16,03
Black	8.879	16,60	8.469	15,77	17.348	16,19
Brown	36.216	67,73	36.434	67,84	72.650	67,78
Total	53.473	100,0	53.706	100,0	107.17	100,0

Source: 1950 Census - Population

The population of Caxias in 2014 was 160,291 (IBGE 2014), with an urban population of 122,254 and a rural population of 38,037. It can be seen, therefore, that this population preferentially lives in the urban center, given that 76.27% of the population is located there, so Caxias is a municipality where the rural environment absorbs only 23.73% of its population.

> *Urbanization Rate*

Table 2 - Degree of urbanization 1970-2010

DEGREE OF URBANIZATION (1970-2010)

Population (1970)	**Caxias**	**Maranhão**	**Brazil**
TOTAL	88.334	3.037.135	93.139.037
Urban	30.735	771.790	52.084.964
Rural	57.599	2.265.345	41.054.053
Degree of urbanization	65,21%	25,40%	55,90%
Population (1980)	**Caxias**	**Maranhão**	**Brazil**
TOTAL	125.606	3.996.404	119.001.037
Urban	56.668	1.255.156	52.084.964

Rural	68.938	2.741.248	41.054.053
Degree of urbanization	45,12%	31,4%	55,9%
Population (1991)	**Caxias**	**Maranhão**	**Brazil**
TOTAL	131.345	4.930.253	146.825.475
Urban	84.331	1.972.421	110.990.990
Rural	47.014	2.957.832	35.834.485
Degree of urbanization	64,21%	40%	75,6%
Population (2000)	**Caxias**	**Maranhão**	**Brazil**
TOTAL	139.756	5.638.361	169.544.443
Urban	103.485	3.355.577	137.697.439
Rural	36.271	2.282.804	31.847.004
Degree of urbanization	74,05%	59,5%	81,2%
Population (2007)	**Caxias**	**Maranhão**	**Brazil**
TOTAL	143.197	6.118.995	183.987.291
Urban	108.542	3.642.638	149.029.706
Rural	34.655	2.476.357	34.957.585
Degree of urbanization	75,82%	59,53%	81%
Population (2010)	**Caxias**	**Maranhão**	**Brazil**
TOTAL	155.129	6.574.789	189.755.799
Urban	118.534	4.147.149	160.925.792
Rural	36.595	2.427.640	28.830.007
Degree of urbanization	76,41%	63,08%	84,81%

Source: IBGE

The process of urbanization in Caxi was essentially based on the rural exodus. Rural-urban migration has multiple causes, the main ones being the loss of work in the agricultural sector due to a concentrated land structure, resulting in a lack of land for the majority of rural workers.

The urbanization rate rose from 13.46% in the 1950s to 45.12 in 1980 and continued to grow. In 2000 it reached 74.05%, with an urban population of around 103,485 inhabitants.

1.4 Factors that contributed to the rise in temperature

1.4.1 Evolution of the motorization rate in Caxias

In December 2013, Caxias' motorization rate (the ratio between the vehicle fleet and the population) was 22.77 vehicles per 100 inhabitants. In June 2014, the motorization rate rose to 26.57. In the period 200 to 2012, the motorization rate went from 35.53 vehicles per 1000 inhabitants to 227.67. During this period, the motorization rate varied by 540.78% and the vehicle fleet increased by 665.67%.

As can be seen from the table below, the periods of greatest increase in the motorization rate were from 2011 to 2013, when it jumped from 173.60 in 2010 to 253.51 in 2013, when it jumped from 173.60 in 2010 to 253.51 in 2013, passing 227.67 in 2012. In this period, the TM had a percentage variation of 46.03%, while the vehicle fleet increased by 50.05% and the population by 2.75% in the period.

Table 3 - Motorization rate in Caxias from 200 to 2013

Year	TM	Year	TM
2000	35,53	2007	112,18
2001	44,39	2008	130,13
2002	52,36	2009	149,83
2003	58,53	2010	173,60
2004	65,10	2011	206,02
2005	76,83	2012	227,67
2006	92,54	2013	253,51

Source: IBGE

From the table above, we can see that the Motorization Rate (MR) per 1000 inhabitants in Caxias was 345.53 vehicles per 1000 inhabitants in 2000. In 2013, it rose to 253.51 vehicles per 1,000 inhabitants.

1.4.2 Industry

Caxias has as its economic activity the construction industry and the processing industry (brewing beer and draft beer). The manufacturing sector was the economic activity whose reduction in participation was due to the decrease in the production of beer and soft drinks.

Between 2002 and 2011, the industry segment grew by 51.89%, equivalent to an average annualized rate of 3.87%. (LETRAS, 2015)

1.5 Urban Climate

All surfaces emit thermal energy according to Planck's Law, which states that the maximum energy that can be emitted from a perfect surface, called a blackbody, depends on the fourth power of the surface temperature: Energy emitted from a blackbody = σT^4

Where: σ is the Stefan-Blotzmann constant (5.67x 10 W/m K^{-824}). No real surface is a black body, so for real surfaces:

Real energy emitted= $\varepsilon\sigma T^4$

Where: ε is the surface emittance, a value between 0 and 1. The emittance of most materials present on Earth ranges from 0.8 to 0.95, but the emittance of uncoated metals can vary from 0.2 to 0.6, depending on the finish of the metal surface.

The urban climate is a complex system, unique to the city. It is the site of profound changes in atmospheric parameters, including air circulation, turbulence and dispersion, albedo and heat storage, evapotranspiration and surface energy balance (TAHA, 1997; ARNFIELD, 2003; KANDA, 2006; LUCENA, 2013). The heat island is the main manifestation of the urban climate and one of the main environmental problems of the 21st century.

This climate is a modification of the local climate which, in relation to the climatic conditions of the surrounding areas, has greater amounts of heat and changes in the composition of the atmosphere, ventilation and humidity, as discussed above.

Poorly planned cities are the ones that suffer the most from the typical problem of urban climate. This situation is described by (SOUZA, 2004; COSTA et al., 2010) "Poorly planned cities suffer serious problems with the phenomenon called heat island, which occurs in places with an excessive amount of buildings and a low concentration of green areas where the temperature is higher than in neighboring non-urbanized areas".

A classic example of a problem typical of cities is the difficulty of obtaining pleasant thermal sensations, which has led the population to resort to artificial forms of heating/cooling, generating an increase in the demand for electricity. In Brazil, the study of thermal comfort by Urban Climatology has confirmed this and other issues in cities with different climatic configurations (SOUZA, 2013; SOUZA et al., 2012).

All of this is because the city's soil, previously covered by vegetation, is being replaced by materials such as asphalt and concrete to promote the expansion of the road network, and this brings to the city an increase in the rate of soil sealing. This waterproofing significantly reduces relative humidity and evaporation due to the low water absorption capacity of these materials.

Traditional asphalt sidewalks tend to be black or grey in color, with solar reflectance values of between 5-10% when new. Asphalt sidewalks become lighter and more reflective with age, however, their relatively dark color tends to maintain their temperature at 65% or more under the summer sun. This condition proves the importance of green areas for the behavior and balance of the urban climate.

1.6 Heat Island

The urban heat island (UHI) is the heat characteristic of urban areas when compared to their (non-urbanized) surroundings and generally refers to the increase in air temperature, but can also refer to the relative heat of the surface or surface materials. In this way, the heat island is a measure of difference (VOOGT, 2002; LUCENA, 2013), which is not evident from an absolute measure of temperature, unduly used heat island phenomenon is necessary to avoid inaccuracies in spatial analysis of the urban climate.

According to Lombardo, the name urban heat island comes from the fact that a city has a very high heat flow in its center, while in its surroundings it is much lower than in the city center.

There are two main reasons for this heating, the first being that most building materials are impermeable and watertight, so there is no moisture available to dissipate the sun's heat. The second is that the combination of dark building materials and pitched floors absorbs and stores more solar energy. Anthropogenic or man-made heat (e.g. burning, deforestation, etc.), lower wind speeds and air pollution in urban areas also contribute to the formation of heat islands.

Consequently, the dynamics of occupation, the prevailing way of life, the different materials used in construction and the regional climate characterization are necessary assumptions. According to GARTLAND (2010), heat islands have five common characteristics:

Air temperatures are high as a result of the heating of urban surfaces, since artificial surfaces absorb more heat from the sun than natural vegetation. These differences in air and surface temperatures are enhanced when the day is calm and clear and are weaker on cloudy and windy days, since more solar energy is captured on clear days, and milder winds remove heat more slowly.

Table 4 - Main urban and suburban characteristics important for the formation of heat islands and their effects on the energy balance over the earth's surface.

Characteristics that contribute to the formation of heat islands	**Effects on the energy balance**
Lack of vegetation	Reduces evaporation
Widespread use of waterproof surfaces	Reduces evaporation
Greater thermal diffusivity of urban materials	Increases heat storage
Low solar reflectance of urban materials	Increases radiation balance
Urban geometries that trap heat	Increases radiation balance
Urban geometries that reduce wind speeds	Reduces convection

Rising pollution levels	Increases radiation balance
Increased energy use	Increases anthropogenic heat

Source: GARTLAND, 2010

1.7 Techniques used to determine heat islands

There are five basic methods used to measure the effects of urbanization on urban climates: fixed stations, mobile transects, remote sensing, vertical sensing and energy balances: Fixed stations, Mobile transects, Remote sensing, Vertical sensing and Energy balances.

1.7.1 Mobile Transects

There are usually too few fixed stations available in the right locations around a city to provide a clear two-dimensional picture of the heat island. Setting up temporary stations to collect data at additional fixed locations can be difficult and costly. Using a mobile transect is a cost-effective way of studying a heat island in urban and suburban areas and their rural surroundings.

A mobile transect involves traveling a predetermined route through a region, stopping at representative locations to take measurements using only one type of meteorological instrumentation. Transportation methods vary. In smaller study areas, researchers have walked or cycled to measurement sites (Spronken Smith and Oke, 1998). In larger areas, researchers used public transportation or cars (Chandler, 1960; Hutcheon et al., 1967; Yamashita, 1996; Stewart, 2000). (GARTLAND, 2010)

Mobile transects can be used at any time of the day or night, although this sometimes depends on traffic conditions. There are some disadvantages to using mobile transect methods to measure heat islands. One of them is the impossibility of obtaining simultaneous measurements in different locations. It is possible to use two or more sets of mobile measuring instruments at the same time, but this would at least double the equipment costs. Most transects can be carried out in less than an hour, but conditions can vary significantly during this time. It is also important to allow the equipment sufficient time to come into equilibrium with its surroundings before taking the measurement.

1.7.2 Remote Sensing

Radioactive energy is transmitted through electromagnetic particles, called photons, which basically act like waves. It is classified according to the lengths of the waves, as shown in the figure.

Figure 3 - Spectrum visible to the human eye

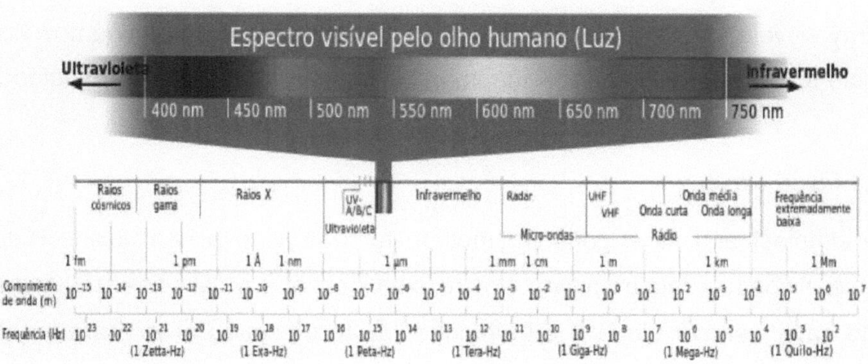

Source: Luz, 2016

The wavelengths that interest us for remote sensing of surface temperatures are in the thermal band, and include some ultraviolet waves and all visible and infrared waves (wavelengths of around 10^{-7} to 10^{-4} m, or 0.1mm to 100 mm). Remote sensing is used to measure temperatures and other characteristics of surfaces, such as roofs, sidewalks, vegetation and bare soil, by means of the energy reflected and emitted from them. Special equipment on aircraft or satellites is used to take pictures of the visible and invisible energy that radiates from cities and their surroundings. The heat island in Buffalo, New York, in 1978, one of the first times satellite images were used to map terrestrial temperatures.

The advantage of using remote sensing is its power to visualize temperatures over large areas. However, remote sensing only shows a panoramic view of urban temperatures, leaving out wall and vegetation temperatures and temperatures under trees. These vertical and shaded surfaces are just as important for an urban heat island as the surfaces seen from above. Some work has been carried out in Vancouver, British Columbia, to add this vertical information to the data collected via remote sensing to generate a truly three-dimensional, or "complete" surface temperature (Voogt and Oke, 1997; GARTLAND, 2010).

Remote sensing instruments usually take measurements at five different energy wavelengths. These measurements can be used to determine how hot or cold surfaces are, and show the image of the surface and how reflected it is.

1.7.3 Fixed stations

The simplest and most common method for analyzing a heat island is to compare weather data from two or more fixed locations. Most cities around the world have weather stations with information accumulated over years on air temperatures, wind speeds, cloudiness, humidity and precipitation levels.

Data from fixed stations has most commonly been used in three different ways: (1) comparing data from a single pair of weather stations, one urban and the other rural; (2) studying data from multiple stations in order to find two-dimensional and regional impacts and (3) investigating a large number of historical data to assess heat island trends over time as a region is urbanized.

If a single pair of stations (one rural and one urban) is used to assess the heat island, it is important to choose the sites to be analyzed carefully. The urban site should be as close as possible to the typical building practices of the region, and the rural site should represent as closely as possible the natural terrain that existed around it before urbanization.

If the region is not well represented by the existing stations, additional fixed stations can be installed temporarily in areas where there are no stations, to collect data at important locations.

An additional weather station should include at least measurements of temperature, humidity, wind speed and direction.

1.7.4 Energy Balance

This balance is based on the first law of thermodynamics, which says that energy is never lost. This means that all the energy absorbed by the surface through radiation or anthropogenic heat goes somewhere. All this energy will either heat the air above the surface or be evaporated with the moisture or stored in the materials in the form of heat. The equation is as follows:

CONVECTION + EVAPORATION + HEAT STORAGE = ANTHROPOGENIC HEAT + RADIATION BALANCE

Convection is the energy that is transferred from a solid surface to a fluid (liquid or gas), in this case from the earth's surface to the air above it. Convection increases when winds are high, when there is turbulence in the air above rougher surfaces and when the temperature differences between the surfaces and the air are greater.

Evaporation is the energy transferred from the earth's surface in the form of water vapor. Water present in moist soils or wet surfaces turns into vapor when it is heated by the sun or other sources. The water vapor then rises into the atmosphere, taking solar energy with it.

Heat storage depends on two material properties: thermal conductivity and heat capacity. Materials with higher thermal conductivity are better able to direct heat to their interior. Materials with a high heat capacity are able to store more heat in their masses. As more heat is stored, the temperature of the material increases.

Radiation balance covers four different radiation processes that take place on the Earth's surface: Radiation balance=global solar radiation-reflected solar radiation+atmospheric radiation-surface radiation. Global solar radiation represents the amount of energy radiated by the sun. This can obviously vary depending on the season, the time of day (zero at night), cloudiness and the level of atmospheric pollution.

Reflected solar radiation is the amount of energy that is reflected from a surface, according to the reflectance of the material. Materials that have a high solar reflectance, such as white, glossy roofing materials, reflect most of the radiation that falls on them, while dark, asphalt-covered surfaces absorb most of the solar radiation.

Atmospheric radiation is the heat emitted by particles in the atmosphere, such as droplets of water vapor, clouds, pollution and dust. The hotter the atmosphere and the more particles present, the more energy will be emitted.

Surface radiation is the heat radiated from a surface. This term is highly dependent on the temperatures of the surfaces and their surroundings. A relatively warmer surface radiates more energy In combination with humidity, building materials have properties that tend to exacerbate heat island problems. Two material properties are important for heat storage: thermal conductivity and heat capacity. Materials with high thermal conductivity tend to conduct heat to their interior. Materials with a high heat capacity are able to store more heat in their volume (GARTLAND, 2010).

Rural areas tend to be made up of less thermally diffusive materials, whereas urban areas are more diffusive. This increases the storage of heat during the day and its slow release at night. Thermal diffusivity can be found by dividing a material's thermal conductivity by its heat capacity.

Energy balance experiments use a lot of equipment to measure the energy flowing to and from surfaces. Four different types of measurements can be made to evaluate the radiation balance term: - **Global solar radiation (short waves):** uses pyranometer or albedometer - **Reflected solar radiation (short waves):** uses pyranometer or albedometer

- **Atmospheric radiation (long waves):** uses a radiometer.
- **Surface radiation (long waves):** uses pygeometer

1.8 Trees as tools for mitigating heat islands

Trees and vegetation reduce heat islands in two ways. Firstly, they shade buildings, sidewalks and people, protecting them from the sun. This keeps surfaces cooler, reduces the heat that is transferred to the air above and reduces the energy consumption of the buildings below them. Shading trees also keep people cooler and more comfortable, reduce the risk of sunstroke and protect them from ultraviolet rays.

Secondly, during the process of photosynthesis, trees and vegetation use a process called evapotranspiration to keep them cool. Plants use solar energy to evaporate water, preventing this energy from being used to heat the city. Air temperatures around and downwind of well-vegetated areas are cooler because of evapotranspiration.

Trees and vegetation absorb water through their roots and emit vapor through their leaves. This process, called evapotranspiration, takes heat from the air to evaporate water. Several studies (HUANG et al., 1990; KURN et al., 1994; GARTLAND, 2010) have found that:

Peak temperatures in wooded areas were 5°C cooler than in open areas (without trees). The air above irrigated agricultural fields was 3°C cooler than the air above land without vegetation. Suburban areas with mature trees were 2-3 °C cooler than new suburbs without trees. Air temperatures over grassy sports fields were 1-2 °C cooler than the air over parking lots.

Plants absorb CO_2 from the atmosphere during the process of photosynthesis, storing carbon for growth and emitting oxygen themselves, trees and vegetation are a vital part of our ecosystem.

2 OBJECTIVES

2.1. General

> Describe the heat island phenomenon in the urban area of Caxias Maranhao using the transect method.

2.2. Specifics

> Comparing temperatures in the center of Caxias with neighborhoods far from the center.

> Measure the temperature during the day (morning and afternoon).

3 METHODOLOGICAL PROCEDURES

3.1 Study Area

Caxias, according to the IBGE, has a territorial extension of 5,197.32 Km^2 ; it corresponds to approximately 1.55% of the territory of Maranhão, of 331,983.293 Km^2 (an area the size of Germany). It has a population density of 30.99 inhabitants/km^2 and an urbanization rate of 76.27%, occupying a prominent place in Maranhão's urban network. It is the fifth most populous municipality in the state (Sâo Luis, Imperatriz, Sâo José de Ribamar and Timon) (LETRAS, 2015). The municipality of Caxias belongs to the Mesoregion Leste Maranhense and the Microregion of Caxias (Caxias, Buriti Bravo, Matoes, Parnarama, Sâo Joâo do Sóter and Timon), with coordinates of 4°51'32"S and 43°21'2"O. Its estimated population in 2013 was 159,396. The population of the urban area, according to the 2010 Census: 104,967 inhabitants. (LETRAS,2015)

3.2 Criteria for choosing research points

With a total of 35 neighborhoods, Caxias was chosen as the area for our climate study. Of these 35 neighborhoods, only five were initially chosen for testing, due to the research methodology being transects. In these five neighborhoods, temperature and humidity data was collected using a thermo-hygrometer, in the morning and afternoon, always within a period of one hour.

The month of June was chosen for our surveys, this time a total of nine neighborhoods were chosen: Salobro, Campo de Belém, Parque da Cidade, Centro, Volta Redonda, Cohab, Praça da Chapada and the Shopping area.

For this research, two thermohygrometers were already used to measure the temperature of the site and the humidity of the air. A laser thermometer was also used to measure the temperature of the soil. The measurements had to be taken within one hour of each other, in the morning and afternoon shifts.

The collection days were Saturdays, Sundays and Mondays. In the morning, measurements began at 9.5 a.m. and ended at 9.45 a.m., and in the afternoon they began at 2.5 p.m. and ended at 2.45 p.m., always with a 10-minute interval from one neighborhood to the next.

These nine points were strategically chosen using Google Earth, where the spatialization of the city was taken into account. We tried to choose points where we could expand, and at the same time cover as much of the city of Caxias as possible, even though there were few collection points. All this can be seen in the following image, which was taken after the points had been chosen.

Figure 4 - Observed data collection points

Source: Google Earth

3.3 Research Equipment

The equipment used for the measurements was: the thermo-hygrometer, which collected data on temperatures ranging from 0 °C to 50 °C / 32 °F to 122 °F, and humidities ranging from 20% to 99%, with a resolution of 1 °C. And the laser thermometer that measured soil temperature, which has a temperature scale ranging from -60 °C to 500 °C, a resolution of 0.1 °C, a focal length of 12:1, an infrared sensor with a laser sight.

Figure 5-Thermohygrometer used in the measurements.

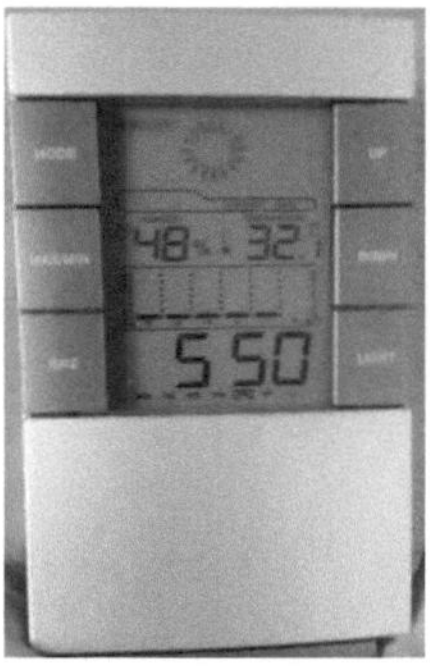

Figure 6 - Laser thermometer used for measurements.

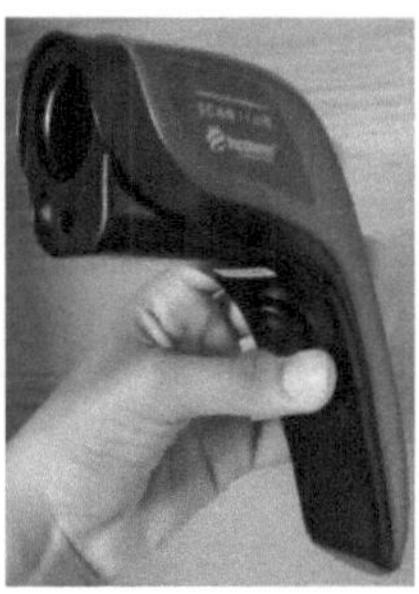

3.4 Characterization of data collection

In the Salobro neighborhood, the collection was done in Rua Getúlio Vargas, in Campo de Belém it was done in Avenida Valter Brito, Parque da Cidade, Centro next to Praça da Matriz, in Morro do Alecrim the collection was done in Praça Duque de Caxias, in Praça da Chapada, in Cohab the collection was done in Avenida 2, in Volta Redonda the analysis was done in Avenida da Coheb and in the Shopping Mall.

Figure 7 - Image of Rua Getúlio Vargas, Bairro Salobro, Caxias-MA.

Source: StreetView - Google Maps

Figure 8 - Image of Avenida Valter Brito, Campo de Belém, Caxias-MA.

Source: Street View - Google Maps

Figure 9 - City park image, Caxias-MA.

Source: Street View - Google Maps

Figure 10 - Image of the Center, Caxias-MA.

Source: Street View - Google Maps

Figure 11 - Image of Morro no Alecrim, highlighting Duque de Caxias Square, Caxias-MA.

Source: Street View - Google Maps

Figure 12 - Chapada Square, Caxias-MA.

Source: Street View - Google Maps

Figure13 - Shopping mall, Caxias-MA.

Source: Street View - Google Maps

Figure 14 - Avenida da Coheb, Volta Redonda, Caxias-MA.

Source: Street View - Google Maps

Figure 15 - Avenue 2 in Cohab, Caxias-MA

Source: Street View - Google Maps

3.4.1 **Collection period**

As previously mentioned, these measurements were taken on Saturdays, Sundays and Mondays, in the morning from 9.05 a.m. to 9.45 a.m. and in the afternoon from 2.05 p.m. to 2.45 p.m. These days were chosen because they had varying flows of people and vehicles in the area, Sunday and Monday were the main days with this difference.

3.4.2 Collection technique

For the research, we used the *Mobile Transect* technique described above, where we pre-determined a strategic route that allowed us to reach the sites before the ten minutes had elapsed, always collecting data in the shade in order to obtain the most accurate measurements. We used thermohygrometers of the same brand, which had to be left in the same environment for more than an hour to calibrate before going into the field, and the clocks had to be adjusted so that there wouldn't be any problems due to the time change.

All the data collected was recorded in a notebook with the following information: temperature, humidity, time, observation (whether it was windy at the time, rain, sun, sun between clouds, cloudy, clear sky, etc.) and then the soil temperature was added to the notes. After this data was entered into tables, initially

Microsoft Word software was used, giving a total of 18 tables, then Microsoft Office Excel was used, with this software it was possible to do all the statistical analysis containing the information needed to carry out this work, all the tables are attached in the appendix.

4 RESULTS AND DISCUSSION

4.1. Temperature data for the morning shift

The neighborhoods chosen have something very intrinsic about them, which is why they were chosen for the analysis described below. After nine days of surveys in the morning and afternoon on June 4th, 5th, 6th, 11th, 12th, 13th, 18th, 19th and 20th, all the data was entered into tables.

All the graphs are analyzed by the regression line where on the ordinate axis (y) we have our explored data and on the abscissa axis (x) the dependent variables, the R^2 is the proportion of the variability of y that is explained by the regression line model, where R^2 can be less than or equal to zero or less than or equal to one.

Table 5 - Temperatures in the research areas in the city of Caxias-MA, in the morning.

TEMPERATURE (°C)

JUNE IN THE MORNING (09:05 TO 09:45)

Salobro	**Belém Field**	**City Park**	**Volta Redonda**	**Rosemary Hill**	**Center**	**Chapada Square**	**Shopping**	**Cohab**
29,20	30,30	31,70	32,70	32,40	32,00	32,50	32,80	33,10
27,40	28,90	30,40	30,40	31,30	30,70	31,00	30,80	31,60
26,30	28,70	31,00	29,90	30,60	30,30	30,70	29,50	30,20
27,30	28,40	29,10	34,50	30,40	30,40	30,10	32,60	33,20
29,70	33,40	35,10	30,10	32,30	38,80	31,80	31,10	30,90
26,50	28,10	30,20	33,00	31,20	30,60	31,40	30,90	31,80
26,10	28,10	30,90	28,80	30,40	30,70	32,40	30,80	31,20
25,70	27,70	29,20	29,10	30,40	29,90	32,40	29,80	31,80
27,30	27,70	29,00	29,60	30,90	30,20	31,60	30,80	31,80
27,3	**29,0**	**30,7**	**30,9**	**31,1**	**31,5**	**31,5**	**31,0**	**31,7**

Source: REIS, 2016

Graph 1 - **Analysis of** morning shift temperatures in the city of Caxias, MA.

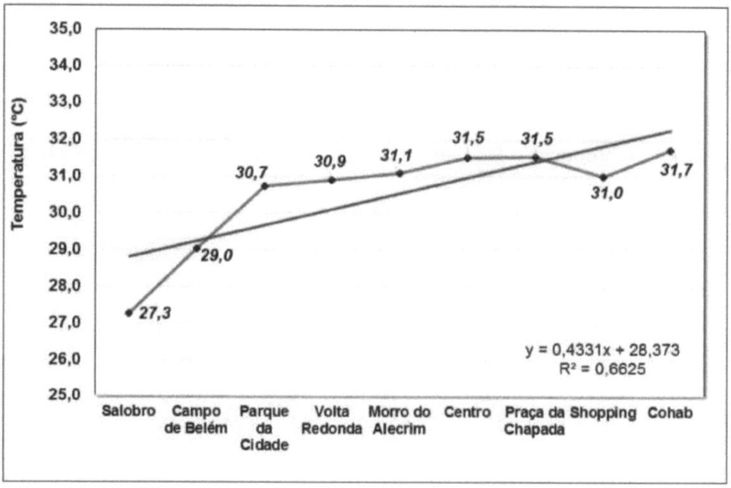

Source: REIS, 2016

Using a thermohygrometer, the temperature and humidity of each neighborhood were collected. After this, they were put into a table, from which the arithmetic average of the temperatures of each neighborhood was extracted in the Excel program and a comparative graph of the temperature of each neighborhood was made. These averages are highlighted in bold in the last line of the table and the results can be seen in the graph.

Graph 1 shows a variation in temperature from one neighborhood to another, where our results can be explained by the regression line (red line), where we observe a positive regression line, which shows an increase in temperature when we move from a less urban environment to a more urban environment, the highest temperature is in the Cohab district, comparing Cohab-Centro the difference in temperature is only 0.2 °C, comparing Cohab-Salobro the difference in temperature between the two districts was 4.4 °C, with this graph alone it would be possible to determine the existence of a heat island in the city of Caxias, but we will now show the data for the afternoon period.

4.2. Temperature data for the afternoon shift

Table 6 - Temperatures in the research areas in the city of Caxias - MA, in the morning.

TEMPERATURE (°C)								
JUNE IN THE AFTERNOON (14:05 TO 14:45)								
Salobro	**Field of**	**Park**	**Volta Redonda**	**Rosemary Hill**	**Center**	**Chapada Square**	**Shopping**	**Cohab**
32,1	33,5	34,5	36,3	34,3	34,1	34,9	34,0	37,9
35,8	37,4	37,5	34,5	37,9	38,8	35,0	34,0	34,8
30,1	31,5	32,9	36,5	33,7	33,1	36,3	36,0	36,8
30,7	32,5	32,1	34,9	33,6	32,6	32,7	35,2	34,1
31,5	33,3	33,9	33,2	36,1	34,9	34,2	35,4	33,6
31,1	33,9	35,0	33,7	34,6	34,6	34,8	35,2	35,6
34,2	34,8	35,4	34,5	35,6	35,6	35,4	37,2	35,2
31,0	33,6	33,7	34,6	34,3	33,8	34,9	35,9	36,4
31,7	33,2	34,1	33,6	34,7	34,4	35,7	35,9	34,8
32,0	**33,7**	**34,3**	**34,6**	**35,0**	**34,7**	**34,9**	**35,4**	**35,5**

Source: REIS, 2016

Graph 2 - Analysis of afternoon temperatures in the city of Caxias, MA.

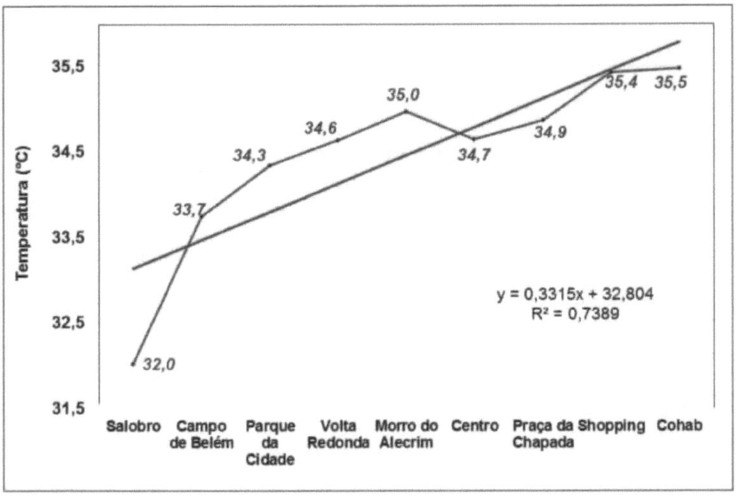

Source: REIS, 2016

The blue line corresponds to the temperature of each neighborhood, it can be seen that there is a sudden increase in temperature, based on the area where the highest temperature value occurred in the transect and observing the positive regression line (red line) we note that it was Cohab in the afternoon, the thermal difference between the other neighborhoods was established and we obtained: Salobro-Cohab 3.5 °C, Campo de Belém-Cohab 1.8 °C, Parque da Cidade-Cohab 1.2 °C, Volta Redonda-Cohab 0.9 °C, Morro do Alecrim-Cohab 0.5 °C, Centro-Cohab 0.8 °C, Praça da Chapada-Centro 0.6 °C, Shopping- Cohab 0.1 °C.

Heat islands are classified as being of low magnitude, when the differences between the points oscillate between 0 °C and 2 °C, medium magnitude between 2 °C and 4 °C, strong between 4 °C and 6 °C and very strong when the differences are greater than 6 °C (GARCIA, 1996; AMORIM, 2005; OLIVEIRA 2012).

The greatest thermal difference found in each transect corresponded to the maximum intensity of the heat island in that transect. The maximum intensity of the heat island in the city was obtained by considering the difference between the maximum value in the Cohab neighborhood and the minimum value in the Salobro neighborhood, which showed a difference of 3.5 °C, thus proving that there are heat islands in the city of Caxias of medium magnitude.

The Salobro neighborhood is far from the center and has cobblestones on the street where the analysis is carried out, a factor that was decisive in the choice of the neighborhood for the research, which had the lowest temperature, even though it is a neighborhood with few trees; Campo de Belém was the second neighborhood with the highest temperature, with cobblestones and a considerable number of trees that are not enough to improve the thermal sensation of the place.

The City Park comes in third, has a ground cover and was a place that intrigued during the research, as it was where the lowest temperature was expected because it was a well-wooded area compared to the other places, after analysis it was found that there are several factors for this increase in temperature, in addition to the paving that the place has, they are: a court made of concrete, several buildings around, apart from the problem of deforestation, in addition to these factors there is the process of evapotranspiration which is the way in which water from the earth's surface passes into the atmosphere in the state of vapor, playing a very important role in the hydrological cycle in global terms, this process involves the evaporation of water from free water surfaces (rivers, lakes, dams, the ocean, etc.), from the soil and from the vegetation.), soils and wet vegetation (which was intercepted during the rain) and plant transpiration. The same happens with the Itapecuru River, which is very close to the site, causing a significant increase in temperature.

Volta Redonda came in fourth. It's a neighborhood far from the center, with wider streets, some of which are still paved, few buildings and little vegetation.

The Center was in fifth place in terms of temperature, it was our main research area with little vegetation, ground cover, flow of vehicles and people, a lot of exposed wiring, very close buildings and narrow streets that make it difficult to pass through, the analysis site was very close to Praça da Matriz which has several trees and the ground without cover which influenced the temperature of the place making it more pleasant compared to the others.

Praça da Chapada came in sixth place, as it has a large flow of people and vehicles. As the analysis site has to be in an ideal environment and without too many influences, the data was collected in the shade in the square, which has a little vegetation, which mitigates the temperature a little; Morro do Alecrim came seventh, despite the fact that the analysis was carried out in Duque de Caxias square and that there are few trees around it, it has a higher temperature due to its altitude.

The Shopping Mall area came in eighth, as it has almost no trees around it, and it is close to the highway that cuts through the city, where there is a flow of vehicles. What mitigates its temperature is the presence of a marsh, but close to it, which is not enough to control its temperature.

Cohab was the neighborhood with the highest temperature, and is the place where heat islands are most present. Despite being a planned neighborhood, it has no vegetation and is paved throughout, which has been a determining factor in this increase, as can be seen in the graph above.

4.3. Relative humidity data for the morning shift

Humidity is another decisive factor in the thermal sensation of a place, which is the amount of water vapor contained in the atmosphere in relation to the maximum amount it could hold at the same temperature (saturation point). During the long dry periods at the end of winter, the humidity in the air drops a lot and is higher on hot summer days because of the evaporation that takes place after rain showers.

In the morning, the humidity is always above 60%, as can be seen below:

Table 7 - Relative humidity in the research areas of the city of Caxias - MA, in the morning.

HUMIDITY (%)

JUNE IN THE MORNING (09:05 TO 09:45)

Salobr o	Belém Field	Park	Volta Redonda	Rosemary Hill	Center	Chapada Square	Shopping	Cohab
80,0	80,0	77,0	72,0	66,0	71,0	65,0	64,0	60,0
80,0	78,0	71,0	68,0	66,0	68,0	69,0	69,0	67,0
81,0	73,0	66,0	72,0	65,0	69,0	62,0	69,0	65,0
80,0	79,0	78,0	60,0	70,0	73,0	71,0	63,0	60,0
72,0	63,0	56,0	70,0	40,0	48,0	65,0	67,0	64,0
81,0	78,0	71,0	60,0	69,0	69,0	65,0	71,0	62,0
71,0	64,0	52,0	62,0	53,0	57,0	47,0	55,0	49,0
79,0	77,0	71,0	67,0	68,0	69,0	60,0	65,0	60,0
72,0	74,0	72,0	62,0	57,0	67,0	59,0	60,0	54,0
77,3	**74,0**	**67,4**	**65,9**	**61,8**	**65,7**	**62,6**	**64,8**	**60,1**

Source: REIS, 2016

Graph 3 - Analysis of air humidity during the morning shift in the city of Caxias-MA

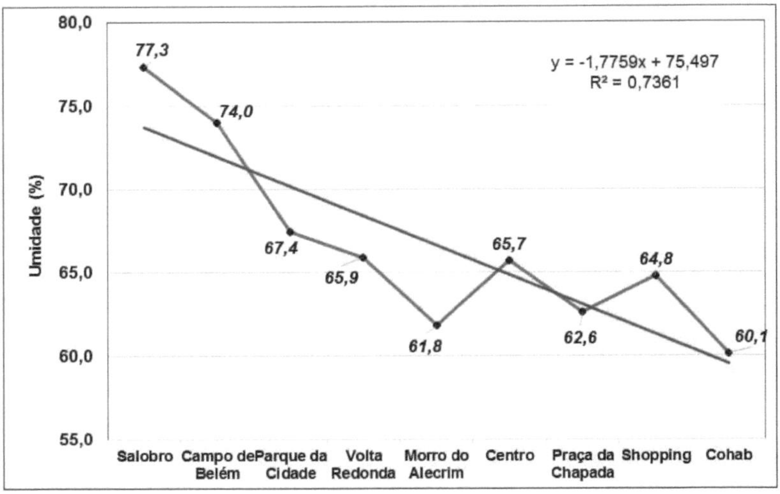

Source: REIS, 2016

Graph 3 shows a negative regression line from the least urban neighborhood with the lowest temperature to the most urban neighborhood with the highest temperature and therefore the lowest humidity. The amount of relative humidity for the neighborhoods during the morning shift shows good results, but the neighborhoods of Salobro and Campo de Belém show an amount above the ideal for human beings, because above 70% there is a risk of mold forming and the spread of certain bacteria. Cohab, despite having the lowest humidity in the morning, still shows a result within the tolerable limit, as can be seen in **graph 3.**

4.4. Relative humidity data for the afternoon shift

The afternoon shift shows a low value, especially in the Center, Cohab and Shopping areas on some collection days, as can be seen in the table below:

Table 8 - Relative humidity in the research areas of the city of Caxias-MA, in the afternoon.

HUMIDITY (%)

JUNE IN THE AFTERNOON (14:05 TO 14:45)

Salobro	Belém Field	City Park	Volta Redonda	Rosemary Hill	Center	Chapada Square	Shopping	Cohab
61	59	52	64	55	50	34	60	52
45	59	45	53	37	38	53	55	50
69	65	52	53	50	54	47	48	45
63	58	58	53	50	57	54	49	52
52	51	51	54	39	45	49	46	50
59	49	46	48	45	45	46	46	45
45	36	36	40	35	34	33	32	33
56	48	45	42	44	45	41	39	35
57	45	45	42	38	48	35	35	35
56,3	**52,2**	**47,8**	**49,9**	**43,7**	**46,2**	**43,6**	**45,6**	**44,1**

Source: REIS, 2016

Graph 4 - Analysis of heat islands with humidity during the afternoon shift in the city of Caxias-MA.

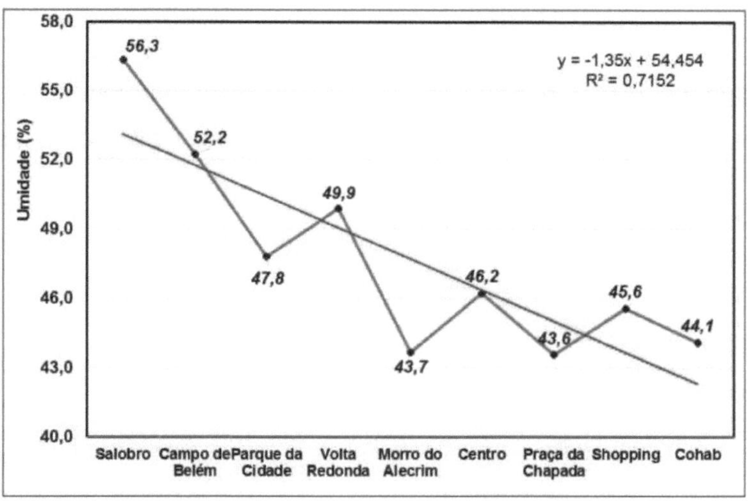

Source: REIS, 2016

Graph 4 shows that Salobro was the neighborhood with the best results in the afternoon, while Praça da Chapada had the lowest relative humidity, although this is within the World Health Organization (WHO) limit, according to which the ideal level for the human body is between 40% and 70%. It is worth noting that on some days the humidity in some neighborhoods was as low as 32%.

According to Varela (2014), low humidity can cause damage to health, as well as hindering the dispersion of polluting gases which aggravate the situation, causing the mucous membranes of the airways to dry out, making the person more vulnerable to asthma attacks and viral and bacterial infections, it also makes the blood denser due to dehydration and favors the appearance of eye problems and allergies.

When the air has a low percentage of humidity (dry air), it tends to obtain water from the elements around it. Conversely, if the air has a high percentage of humidity, it will have the ability to donate water to the elements that are drier.

4.5. Soil temperature data for the morning shift

For the analysis of heat islands in the last week of analysis, the following were used

measurements of the soil at the collection point with the laser thermometer, as most of the points had a common ground cover, these data can be seen in the table below:

Table 9 - Soil temperature in the research areas in the city of Caxias, MA, in the afternoon.

SOIL TEMPERATURE (°C)

JUNE IN THE MORNING (09:05 TO 09:45)

Salobro	Belém Field	City Park	Volta Redonda	Rosemary Hill	Center	Chapada Square	Shopping	Cohab
28,40	34,70	28,30	26,90	29,00	29,60	30,00	24,10	32,00
26,80	31,60	27,70	26,70	30,70	34,60	30,50	28,40	36,20
27,70	33,20	28,30	27,10	29,30	32,70	29,80	28,70	34,10

27,63	33,17	28,10	26,90	29,67	32,30	30,10	27,07	34,10

Source: REIS, 2016

Graph 5 - Analysis of soil temperatures during the morning shift.

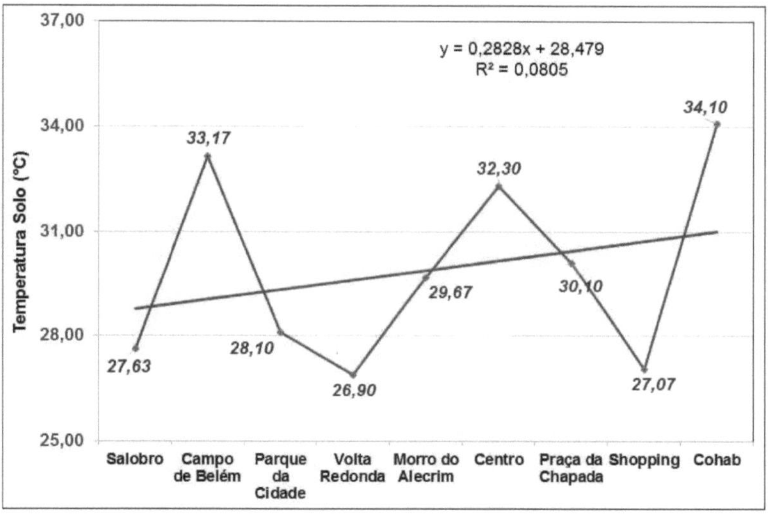

Looking at **graph 5**, we can see that the warmest sidewalks tend to be impermeable and dark in color, with reflectance below 25%. Asphalt sidewalks generally have thermal emissivity values of 85% or more, while the thermal emissivity of concrete sidewalks varies between 70 and 90%. But the effects of thermal emissions are outweighed by the effects of solar reflectance. For dry, impermeable floors, solar reflectance is the factor that most affects their temperature. Since floor colors have the greatest effect on solar reflectance, lighter-colored floors are generally the coolest. You can see in the graph that Cohab leads the way in terms of ground temperature, both in this neighborhood and in the others, with the exception of Salobro, where the ground cover is darkly pigmented, i.e. where solar reflectance is low.

4.6. Soil temperature data for the afternoon shift

Table 10 - Soil temperature in the research areas in the city of Caxias-MA, in the afternoon.

SOIL TEMPERATURE (°C)

MONTH OF JUNE IN THE AFTERNOON (14						**:05 TO 14:45)**		
Salobro	**Belém Field**	**City Park**	**Volta Redonda**	**Rosemary Hill**	**Center**	**Chapada Square**	**Shopping**	**Cohab**
34,10	38,10	42,60	33,70	39,60	33,20	32,90	43,70	34,50
38,40	38,40	40,60	33,30	39,80	34,40	34,70	39,30	34,90
37,60	39,20	40,40	34,50	31,90	34,40	34,40	39,70	34,50
36,7	**38,6**	**41,2**	**33,8**	**37,1**	**34,0**	**34,0**	**40,9**	**34,6**

Source: REIS, 2016

Graph 6 - Soil temperature data, afternoon shift.

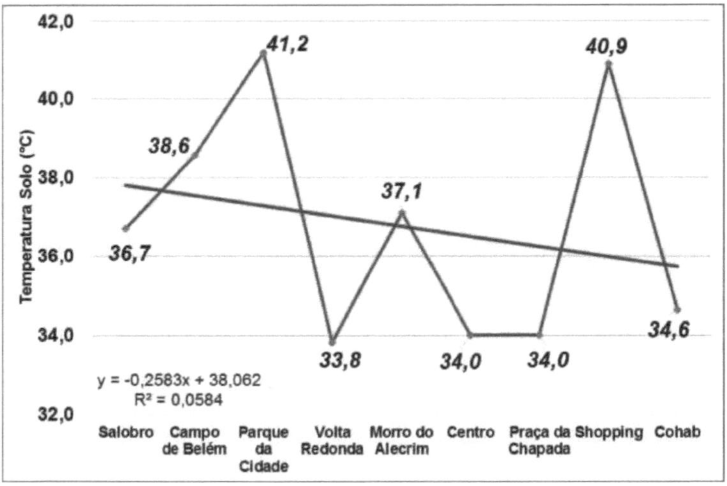

Source: REIS, 2016

The data for Parque da Cidade and the shopping center area has

higher values in terms of temperature as can be seen in **graph 6**. Sidewalks cover between 25 and 50% of the city of Caxias and are the most dominant aspect of our urban environments. The thermal characteristics of sidewalks have a major influence on the formation of heat islands. Asphalt can be heated to 65 °C or more under the summer sun and is the second hottest aspect in the urban landscape, behind traditional roofing materials.

Asphalt sidewalks are generally not cooled, but they can be cooled in various ways. Light pigments or light-colored aggregates (stones in the sidewalk mix) can be added to the asphalt mix. These additives can increase the solar reflectance of sidewalks by up to 30 percentage points and can raise a material's reflectance from 15% to 45%.

Another way to keep floors cool is to make them porous or permeable. This would allow rainwater to accumulate in the material and thus keep the floor cooler over time.

FINAL CONSIDERATIONS

Based on the results of this study, it is possible to conclude that:

- Caxias has medium heat islands with a difference of 3.5 °C from the neighborhood with the lowest temperature (Salobro) to the neighborhood with the highest temperature (Cohab);
- All heat islands (both less intense and more intense) are made up of materials such as asphalt, cement and roof tiles that interfere with the local energy balance and places with little or no green space;
- Possible solutions could come from a more concrete perception of the reality of Brazilian society as a whole and from participatory work between the administration and society;
- In fact, the change in land use and cover (from natural to urbanized with building materials) affects local and regional temperatures;
- Programs to encourage the planting of trees in the city of Caxias should be implemented, and alternatives should be sought for cooler floors and roofs, and people should be made aware of the existence of heat islands;
- Mitigating heat islands holds great promise, not only for making our neighborhoods more comfortable and livable, but also for reducing environmental impacts and construction and maintenance costs in our cities;
- Mitigating heat islands would bring several benefits, among them: improved health, lower temperatures and relative humidity, energy savings and, consequently, a better quality of life for people;
- I end the paper by transcribing a reference by Chandler, one of the forerunners of urban climate studies, in whose conclusions 25 years ago, which remain valid today, he admitted: "the reason for the neglect of the climate issue has been partly the relatively poor communication links that currently exist between Urban Climatology and partly the relatively poor communication links

that currently exist between Climatology and Planning. But in view of the exponential growth of the world's population and the increasing pace of urbanization, it is clear that our cities must, where appropriate, be properly planned in order to optimize the urban environment and avoid a series of structural and functional design flaws. Climate is an essential element in this planning (Chandler, 1976, cited by Gregory, 1992; MONTEIRO, 2003).

BIBLIOGRAPHICAL REFERENCES

ANDRADE, Leidiane do Livramento *et al*, Studies of heat islands in the city of Sâo José dos Campos using the thermal infrared channel of LandSat-5 and the airborne HSS. Available at: <http://www.sjc.sp.gov.br>. Accessed on 11/12/2015

BIAS, Edilson de Souza *et al*, Analysis of the phenomenon of Urban Heat Islands, through the combination of LandSat and Ikonos data. Available at:<http://marte.sid.inpe.br> .Accessed on 11/12/2015

BRITO, Tainà Laeta Felipe de, Caracteristicas da Ilha de Calor Urbana de verão e primavera na baixada de Jacarepaguà. Available at: <http://www.geocart.igeo.ufrj.br>. Accessed on: 02/05/2016

CARVALHO,Dark Lane Rodrigues *et al*, Identification of Urban Heat Islands in the City of Manaus-AM. Available at: <http://www.drs.inpe.br>.Accessed on: 11/12/2015

COSTA,Douglas F. da *et al*, Identification of Heat Islands in the Urban Area of Ilha Solteira-SP through the Use of Geotechnologies. Available at: <http://repositorio.unesp.br>. Accessed on: 28/03/2015

COSTA, Douglas Fernandes da *et al*, Identification of Heat Islands in the Urban Area of Ilha Solteira-SP through the use of geotechnologies. Available at: <http://marte.sid.inpe.br> . Accessed on: 11/12/2015

FRANÇA,LucianaMayla de Aquino *et al*, Analysis of the thermal discomfort caused by the Urban Heat Island in a neighborhood of Recife-PE using Remote Sensing. Available at:< http://www.drs.inpe.br>. Accessed on: 11/12/2015

FILHO, Francisco de Assis Leandro, Numerical Study of the Heat Island in Fortaleza using the RAMS model. Available at: <http://www.uece.br> Accessed on: 02/05/2016

GARTLAND, Lisa. Heat Islands: How to mitigate heat zones in urban areas. 1ª edition. Sâo Paulo: Oficina de Textos, 2010

LEMOS, Jacqueline de Souza, Spatialization of the Heat Island of the Urban Agglomeration of the Metropolitan Region of Curitiba (AU-RMC) in August 2006 from surface thermography. Available at:< http://www.drs.inpe.br>. Accessed on: 11/12/2015

LETRAS, Academia Caxiense de, Cartografias Invisiveis: Saberes e Sentires de Caxias. 1ª edition. Caxias: JM Ltda, 2015

LUZ, Virtual Museum of Light. Available at: <http://museudaluz.com.br. Accessed: 27/07/2016

MONTEIRO, C. A.F.; MENDONÇA, F. Clima urbano. Sâo Paulo: Contexto, 2003

MORAIS, Heliene Ferreira de *et al*, Identification of the phenomenon of Urban Heat Islands in Belo Horizonte-MG. Available at:< http://www.drs.inpe.br>. Accessed on: 11/12/2015

MOREIRA, Elvis Bergue Mariz *et al*, Identification of the phenomenon of Heat Islands in the Urban Area of Recife-PE, through the Thermal Infrared Channel of the LandSat 5 Satellite. Available at: <http://www.drs.inpe.br>. Accessed on: 11/12/2015

OLIVEIRA, Jorge Ricardo Felix de, Heat islands in small towns: the case of Pacatuba-CE. Available at:

<http://www.revistageonorte.ufam.edu.br>.Accessed on: 28/03/2016

SILVA, Ivanir Ortega Rodrigues da, A relação Sociedade-Nature e alguns aspectos sobre planejamento e gestão ambiental no Brasil. Available at:<http://web.ua.es>. Accessed on: 29/03/2016

SOUZA, Débora Moreira de *et al*, Thermal comfort from the perspective of Geographical Climatology. Available at: <http://www.uel.br>. Accessed on: 02/05/2016

SOUZA, Rose Marry dos Santos, Surface temperatures in Guarabira as a heat island phenomenon. Available at:< http://repositorio.unesp.br>. Accessed on: 28/03/2015

TEZA, Clàusio Tavares Viana *et al*, Identification of the Urban Heat Islands phenomenon using ASTER ondemand 08-Kinetic Temperature (III) data: Brazilian metropolises. Available at:<http://marte.sid.inpe.br>. Accessed on: 11/12/2015

VENTURINI, Ana Carolina dos Santos, Cenàrio Urbano: O desenvolvimento e o microlema da cidade. Available at: <http:WWW.bibliotecadigital.ufmg.br> Accessed on: 10/05/2016

APPENDIX

Table of measurements taken during the survey

SATURDAY					
June 04, 2016 (Morning)					
NEIGHBORHOODS	**ALTITUDE**	**TEMPERATURE (°C)**	**HUMIDITY (%)**	**SCHEDULE**	**NOTE**
Salobro	81 m	29,2	80	09:05	Sun
Belém Field	72m	30,3	80	09:15	Sun
City Park	60m	31,7	70	09:25	Sun/Clouds
Center	72m	32	71	09:35	Sun
Rosemary Hill	105m	32,4	68	09:45	Sun/Clouds
Coheb Avenue	90m	32,7	72	09:05	Sun
Shopping	103m	32,8	64	09:15	Sun
Cohab	112m	33,1	60	09:25	Sun
Chapada Square	79m	32,5	65	09:35	Sun/Clouds

Note: Average temperature at the time of measurement 26 °C

SATURDAY					
June 04, 2016 (Afternoon)					
NEIGHBORHOODS	**ALTITUDE**	**TEMPERATURE (°C)**	**HUMIDITY (%)**	**SCHEDULE**	**NOTE**
Salobro	81 m	32,1	61	14:05	Sun/Clouds
Belém Field	72m	33,5	59	14:15	Sun
City Park	60m	34,5	52	14:25	Sun
Center	72m	34,1	50	14:35	Sun/Clouds
Rosemary Hill	105m	34,3	55	14:45	Sun/Clouds
Coheb Avenue	90m	36,3	64	14:05	Sun
Shopping	103m	34	60	14:15	Sun
Cohab	112m	37,9	52	14:25	Sun
Chapada Square	79m	34,9	34	14:35	Sun

Note: Average temperature at the time of measurement 32 °C

SUNDAY					
June 05, 2016 (Morning)					
NEIGHBORHOODS	**ALTITUDE**	**TEMPERATURE (°C)**	**HUMIDITY**	**SCHEDULE**	**NOTE**

			(%)		
Salobro	81 m	27,4	80	09:05	Sun/Clouds
Belém Field	72m	28,9	78	09:15	Sun
City Park	60m	30,4	71	09:25	Sun
Center	72m	30,7	68	09:35	Sun
Rosemary Hill	105m	31,3	66	09:45	Sun
Coheb Avenue	90m	30,4	68	09:05	Sun
Shopping	103m	30,8	69	09:15	Sun/Clouds
Cohab	112m	31,6	67	09:25	Sun/Clouds
Chapada Square	79m	31	69	09:35	Sun

Note: Average temperature at the time of measurement 29 °C

SUNDAY					
June 05, 2016 (Afternoon)					
NEIGHBORHOODS	**ALTITUDE**	**TEMPERATURE (°C)**	**HUMIDITY (%)**	**SCHEDULE**	**NOTE**
Salobro	81 m	35,8	45	14:05	Sun/Clouds
Belém Field	72m	37,4	39	14:15	Sun/Clouds/Wind
City Park	60m	37,5	45	14:25	Sun
Center	72m	38,8	38	14:35	Sun/
Rosemary Hill	105m	37,9	37	14:45	Sun/Wind
Coheb Avenue	90m	34,5	53	14:05	Sun/Clouds
Shopping	103m	34	55	14:15	Sun/Clouds
Cohab	112m	34,8	50	14:25	Sun/Clouds
Chapada Square	79m	35	53	14:35	Sun

Note: Average temperature at the time of measurement 32 °C

MONDAY					
June 06, 2016 (Morning)					
NEIGHBORHOODS	**ALTITUDE**	**TEMPERATURE (°C)**	**HUMIDITY (%)**	**SCHEDULE**	**NOTE**
Salobro	81 m	26,3	81	09:05	Sun
Belém Field	72m	28,7	73	09:15	Sun/Clouds
City Park	60m	31	66	09:25	Sun
Center	72m	30,3	69	09:35	Sun/Clouds
Rosemary Hill	105m	30,6	65	09:45	Sun/Clouds

Coheb Avenue	90m	29,9	72	09:05	Sun
Shopping	103m	29,5	69	09:15	Sun
Cohab	112m	30,2	65	09:25	Sun
Chapada Square	79m	30,7	62	09:35	Sun

Note: Average temperature at the time of measurement 28 °C

MONDAY					
June 06, 2016 (Afternoon)					
NEIGHBORHOODS	**ALTITUDE**	**TEMPERATURE (°C)**	**HUMIDITY (%)**	**SCHEDULE**	**NOTE**
Salobro	81 m	30,1	69	14:05	Sun/Clouds
Belém Field	72m	31,5	65	14:15	Sun/Clouds/Wind
City Park	60m	32,9	52	14:25	Sun
Center	72m	33,1	54	14:35	Sun
Rosemary Hill	105m	33,7	50	14:45	Sun/Wind
Coheb Avenue	90m	36,5	53	14:05	Sun
Shopping	103m	36	48	14:15	Sun
Cohab	112m	36,8	45	14:25	Sun
Chapada Square	79m	36,3	47	14:35	Sun

Note: Average temperature at the time of measurement 33 °C

SATURDAY					
June 11, 2016 (Morning)					
NEIGHBORHOODS	**ALTITUDE**	**TEMPERATURE (°C)**	**HUMIDITY (%)**	**SCHEDULE**	**NOTE**
Salobro	81 m	27,3	80	09:05	Sun/Clouds
Belém Field	72m	28,4	79	09:15	Sun
City Park	60m	29,1	78	09:25	Sun/Clouds
Center	72m	30,4	73	09:35	Sun/Clouds
Rosemary Hill	105m	30,4	70	09:45	Sun/Clouds
Coheb Avenue	90m	34,5	60	09:05	Sun/Clouds
Shopping	103m	32,6	63	09:15	Sun/Clouds
Cohab	112m	33,2	60	09:25	Sun/Youth/Wind
Chapada Square	79m	30,1	71	09:35	Sun

Note: Average temperature at time of measurements **22°**

SATURDAY

June 11, 2016 (Afternoon)					
NEIGHBORHOODS	**ALTITUDE**	**TEMPERATURE (°C)**	**HUMIDITY (%)**	**SCHEDULE**	**NOTE**
Salobro	81 m	30,7	63	14:05	Partially Cloudy
Belém Field	72m	32,5	58	14:15	Partially Cloudy
City Park	60m	32,1	58	14:25	Partially Cloudy
Center	72m	32,6	57	14:35	Partially Cloudy/ Serenading
Rosemary Hill	105m	33,6	50	14:45	Partially Cloudy
Coheb Avenue	90m	34,9	53	14:05	Sun/Clouds
Shopping	103m	35,2	49	14:15	Sun/Youth/ Wind/Snow
Cohab	112m	34,1	52	14:25	Sun/Youth/ Wind/Snow
Chapada Square	79m	32,7	54	14:35	Sun/Youth/ Wind

Note: Average temperature at time of measurements **26°**

SUNDAY					
June 12, 2016 (Morning)					
NEIGHBORHOODS	**ALTITUDE**	**TEMPERATURE (°C)**	**HUMIDITY (%)**	**SCHEDULE**	**NOTE**
Salobro	81 m	29,7	72	09:05	Sun/Clouds
Belém Field	72m	33,4	63	09:15	Sun/Clouds
City Park	60m	35,1	56	09:25	Sun/Clouds
Center	72m	38,8	48	09:35	Sun/Clouds
Rosemary Hill	105m	32,3	40	09:45	Sun/Clouds
Coheb Avenue	90m	30,1	70	09:05	Sun/Clouds
Shopping	103m	31,1	67	09:15	Sun/Wind
Cohab	112m	30,9	64	09:25	Sun/Wind
Chapada Square	79m	31,8	65	09:35	Sun/Wind

Note: Average temperature at time of measurements **26°**

SUNDAY					
June 12, 2016 (Afternoon)					
NEIGHBORHOODS	**ALTITUDE**	**TEMPERATURE (°C)**	**HUMIDITY (%)**	**SCHEDULE**	**NOTE**
Salobro	81 m	31,5	52	14:05	Sun/Clouds
Belém Field	72m	33,3	51	14:15	Sun
City Park	60m	33,9	51	14:25	Sun/Clouds
Center	72m	34,9	45	14:35	Sun
Rosemary Hill	105m	36,1	39	14:45	Sun
Coheb Avenue	90m	33,2	54	14:05	Sun
Shopping	103m	35,4	46	14:15	Sun
Cohab	112m	33,6	50	14:25	Sun
Chapada Square	79m	34,2	49	14:35	Sun

Note: Average temperature at time of measurements **30°**

MONDAY					
June 13, 2016 (Morning)					
NEIGHBORHOODS	**ALTITUDE**	**TEMPERATURE (°C)**	**HUMIDITY (%)**	**SCHEDULE**	**NOTE**
Salobro	81 m	26,5	81	09:05	Sun
Belém Field	72m	28,1	78	09:15	Sun
City Park	60m	30,2	71	09:25	Sun
Center	72m	30,6	69	09:35	Sun/Clouds
Rosemary Hill	105m	31,2	69	09:45	Sun
Coheb Avenue	90m	33	60	09:05	Clear skies/ Wind
Shopping	103m	30,9	71	09:15	Sun
Cohab	112m	31,8	62	09:25	Sun/Wind
Chapada Square	79m	31,4	65	09:35	Clear skies/Wind

Note: Average temperature at the time of measurement **24°**.

MONDAY					
June 13, 2016 (Afternoon)					
NEIGHBORHOODS	**ALTITUDE**	**TEMPERATURE (°C)**	**HUMIDITY (%)**	**SCHEDULE**	**NOTE**
Salobro	81 m	31,1	59	14:05	Sun
Belém Field	72m	33,9	49	14:15	Sun

City Park	60m	35	46	14:25	Sun/Clouds
Center	72m	34,6	45	14:35	Sun/Clouds
Rosemary Hill	105m	34,6	45	14:45	Sun/Clouds
Coheb Avenue	90m	33,7	48	14:05	Clear Skies
Shopping	103m	35,2	46	14:15	Sun/Clouds
Cohab	112m	35,6	45	14:25	Sun/Wind
Chapada Square	79m	34,8	46	14:35	Sun/Wind

Note: Average temperature at time of measurements **30°**

SATURDAY					
June 18, 2016 (Morning)					
NEIGHBORHOODS	**ALTITUDE**	**TEMPERATURE (°C)**	**HUMIDITY (%)**	**SCHEDULE**	**NOTE**
Salobro	81 m	26,1	71	09:05	Sun
Belém Field	72m	28,1	64	09:15	Sun
City Park	60m	30,9	52	09:25	Sun
Center	72m	30,7	57	09:35	Sun
Rosemary Hill	105m	30,4	53	09:45	Sun
Coheb Avenue	90m	28,8	62	09:05	Clear skies/Wind
Shopping	103m	30,8	55	09:15	Clear Skies
Cohab	112m	31,2	49	09:25	Wind/Clear Sky
Chapada Square	79m	32,4	47	09:35	Clear skies/Wind

SATURDAY					
June 18th, 2016 (Afternoon)					
NEIGHBORHOODS	**ALTITUDE**	**TEMPERATURE (°C)**	**HUMIDITY (%)**	**SCHEDULE**	**NOTE**
Salobro	81 m	34,1	45	14:05	Sun
Belém Field	72m	34,8	36	14:15	Sun/Clouds
City Park	60m	35,4	36	14:25	Sun/Clouds
Center	72m	35,6	34	14:35	Sun/Clouds
Rosemary Hill	105m	35,6	35	14:45	Sun
Coheb Avenue	90m	34,5	40	14:05	Sun
Shopping	103m	37,2	32	14:15	Sun
Cohab	112m	35,2	33	14:25	Sun/Wind

Chapada Square	79m	35,4	33	14:35	Sun/Wind

SUNDAY					
June 19, 2016 (Morning)					
NEIGHBORHOODS	**ALTITUDE**	**TEMPERATURE (°C)**	**HUMIDITY (%)**	**SCHEDULE**	**NOTE**
Salobro	81 m	25,7	79	09:05	Clear Skies
Belém Field	72m	27,7	77	09:15	Clear Skies
City Park	60m	29,2	71	09:25	Clear Skies
Center	72m	29,9	69	09:35	Clear Skies
Rosemary Hill	105m	30,4	68	09:45	Clear Skies
Coheb Avenue	90m	29,1	67	09:05	Clear skies/Wind
Shopping	103m	29,8	65	09:15	Clear Sky/Wind
Cohab	112m	31,8	60	09:25	Clear Sky/Wind
Chapada Square	79m	32,4	60	09:35	Clear skies/Wind

SUNDAY					
June 19, 2016 (Afternoon)					
NEIGHBORHOODS	**ALTITUDE**	**TEMPERATURE (°C)**	**HUMIDITY (%)**	**SCHEDULE**	**NOTE**
Salobro	81 m	31	56	14:05	Sun/Clouds
Belém Field	72m	33,6	48	14:15	Sun/Clouds
City Park	60m	33,7	45	14:25	Sun/Clouds
Center	72m	33,8	45	14:35	Sun/Clouds
Rosemary Hill	105m	34,3	44	14:45	Clear Skies
Coheb Avenue	90m	34,6	42	14:05	Sun/Wind
Shopping	103m	35,9	39	14:15	Sun
Cohab	112m	36,4	35	14:25	Sun/Wind
Chapada Square	79m	34,9	41	14:35	Sun/Wind

MONDAY					
June 20, 2016 (Morning)					
NEIGHBORHOODS	**ALTITUDE**	**TEMPERATURE (°C)**	**HUMIDITY (%)**	**SCHEDULE**	**NOTE**

Salobro	81 m	27,3	72	09:05	Clear Skies
Belém Field	72m	27,7	74	09:15	Sun/Clouds
City Park	60m	29	72	09:25	Sun/Clouds
Center	72m	30,2	67	09:35	Sun/Clouds
Rosemary Hill	105m	30,9	56	09:45	Sun
Coheb Avenue	90m	29,6	62	09:05	Clear skies/Wind
Shopping	103m	30,8	60	09:15	Clear Skies
Cohab	112m	31,8	54	09:25	Clear Sky/Wind
Chapada Square	79m	31,6	59	09:35	Clear skies

SATURDAY					
June 20, 2016 (Afternoon)					
NEIGHBORHOODS	**ALTITUDE**	**TEMPERATURE (°C)**	**HUMIDITY (%)**	**SCHEDULE**	**NOTE**
Salobro	81 m	31,7	57	14:05	Sun/Clouds
Belém Field	72m	33,2	45	14:15	Sun/Clouds
City Park	60m	34,1	45	14:25	Sun/Clouds
Center	72m	34,4	48	14:35	Sun/Clouds
Rosemary Hill	105m	34,7	38	14:45	Sun/Clouds
Coheb Avenue	90m	33,6	42	14:05	Sun Wind
Shopping	103m	35,9	35	14:15	Sun Wind
Cohab	112m	34,8	35	14:25	Sun/Wind
Chapada Square	79m	35,7	35	14:35	Sun/Wind

MIX
Papier aus verantwortungsvollen Quellen
Paper from responsible sources
FSC® C105338
FSC
www.fsc.org

Printed by Books on Demand GmbH, Norderstedt / Germany